A Wisley Handbook

Propagation from Cuttings

JIM GARDINER

Cassell

The Royal Horticultural Society

THE ROYAL HORTICULTURAL SOCIETY

Cassell Educational Limited
Wellington House, 125 Strand
London WC2R 0BB
for the Royal Horticultural Society

First published 1997
Reprinted 1997

British Library Cataloguing in Publication Data
A catalogue record for this book is available from the British Library

ISBN 0–304–34421–4

Photographs by Jim Gardiner (pp. 9, 14, 28, 29); RHS/Tim Sandall (pp. 8, 11, 16, 17, 20, 23, 25, 30, 41, 42, 51, 57); Neil Campbell Sharp (p. 49 (right)); Harry Smith Horticultural Photographic Collection (pp. 4, 13, 19, 34, 35, 38, 40, 43, 44, 46, 49 (left), 50, 53, 55, 59, 61, 62, 63)

Line drawings by Mike Shoebridge

Typeset by RGM, The Mews, Birkdale Village, Southport
Printed in Hong Kong by Wing King Tong Co. Ltd.

Cover: Cuttings of euonymus should root within 6–8 weeks.
Photograph by RHS/Tim Sandall
Back cover: A mixed border at The Priory, Kelmerton, features herbaceous perennials, all of which can be propagated by division.
Photograph by Andrew Lawson
p. 1: *Paeonia* 'Duchesse de Nemours', like other herbaceous peonies, should be divided in the spring.
Photograph by Neil Campbell Sharp

Contents

Introduction

There are many ways of raising your own plants, ranging from the simple methods of sowing seeds to the more complicated art of budding and grafting. With the exception of seed sowing (which is fully covered in the Wisley Handbook *Propagation from Seed*), all methods of raising new plants are collectively known by the term vegetative propagation. By this means one part of the plant, primarily a stem, root or leaf, has the ability to initiate and develop a new plant which will be identical to its parent. To propagate successfully requires not only practise, but knowledge of the structure of the plant, how it grows and its environmental requirements. These important points are discussed in the following pages in conjunction with the various techniques that can be employed.

Vegetative propagation can be divided into three main sections:

- Cuttings, including layering and suckering
- Division
- Grafting

In this book, emphasis is placed on raising plants from cuttings and by division. Budding and grafting have been excluded on the grounds that they require considerable skill, and most plants can be raised more easily by an alternative and simpler method. The equipment needed for the various techniques is discussed. Full details are given of how to care for the material through the various stages, from taking the cuttings to planting out.

Micropropagation is another method of raising plants from cuttings. However, it calls for specialist techniques and maintaining conditions that are, for the most part, beyond the scope of the average gardener.

This book will help you to propagate plants vegetatively using simple and reliable techniques. However, at the end of the day, propagation is a practical art and it is practice that will translate this theoretical knowledge into success.

Fuchsias root fastest from softwood cuttings taken in spring

Factors Affecting Success

There are a number of points that affect the successful propagation of plants. As well as the environmental factors and equipment used, the selection and treatment of the plant material are important.

SELECTING SUITABLE MATERIAL

This is probably the most important factor when propagating by cuttings. Appropriate material will sometimes root under inadequate conditions but poor material will rarely root. Only the best forms of a particular plant should be used, and the material must be healthy.

Nutrient levels

The nutrient status of the parent plant will affect the ability of the potential cutting to root. High levels of carbohydrate or starch and low-to-medium nitrogen levels often favour rooting. This means that rank growing, lush and succulent material should be avoided.

Juvenility

The age of the parent plant affects the ease with which cuttings root, especially in the case of difficult plants. Juvenile plants (those that have not yet flowered and set seed) will root readily. However, as soon as a plant has flowered it will start to 'age' and its ability to root will decline. Aging can be arrested by pruning which will encourage strong-growing, often non-flowering, shoots which root fairly readily. It is often difficult to accept that a plant rooted this year is the same age physiologically as a plant raised fom the same stock many years before.

Type of wood selected

On woody plants there is a wide range of material some of which is suitable for one method of propagation but not another. Different rates of success are often shown between cuttings taken from terminal and lateral shoots. Variations in rooting are also found between basal and tip sections of a shoot, cuttings with or without a heel, flowering or non-flowering (vegetative) shoots. However, recommendations are given in the charts on pages 52–64 to help you determine which material gives the best results.

Timing

Cuttings can be taken throughout the year, but the optimum time for successful rooting varies. This variation is probably due more to the physiological state of the cutting than to the calendar date. Timing recommendations are also given in the charts.

TREATMENT OF CUTTINGS

Rooting can be encouraged by using chemicals – rooting hormones and fungicides – and by physical and environmental means.

Rooting hormones

During the 1930s it was discovered that auxins (plant hormones) such as indole-3-acetic acid (IAA) stimulated the production of roots in both stem and leaf cuttings. This chemical is not generally used today because of its unstable nature. However X-naphthalene acetic acid (NAA) and, more generally, indole butyric acid (IBA) are used in commercial preparations. The benefits of using such substances include increased percentages of rooted cuttings, quicker root initiation and general improvement in quality and quantity of the roots. However, rooting hormones do not guarantee success.

As they are needed in very small quantities, rooting hormones are generally mixed with either fine talcum powder or with an organic solvent such as alcohol. Other additives include fungicides that help to prevent basal rots.

Powder formulations are most commonly available due to their ease of application and broad safety margin of use. The active ingredient is generally expressed as a percentage, e.g. 0.1% IBA. As a guide:

- 0.1% IBA (no. 1) is used for softwood cuttings;
- 0.3% IBA (no. 2) is used for greenwood and semi-ripe cuttings;
- 0.8% IBA (no. 3) is used for hardwood and difficult-to-root semi-ripe cuttings.

When powder formulations are not in use they should be stored in a fridge at between 3–5°C (37–40°F).

Liquid formulations are often more effective than powders. They are generally expressed in parts per million (ppm):

The base of a semi-ripe cutting of *Garrya elliptica* is dipped quickly into an alcohol-based hormone rooting formulation

- 500–1,000 ppm for softwood cuttings;
- 1,000–2,500 ppm for greenwood and semi-ripe cuttings;
- 2,500–5,000 ppm for hardwood and difficult-to-root semi-ripe cuttings.

How to apply rooting hormones

As hormones are absorbed through the cut surface, dip the base of the stem to a depth of 1 cm ($\frac{1}{4}$–$\frac{1}{2}$ in). With powders it is advisable to dip the base of the cutting first in water to help adhesion, then to tap off excess powder.

Liquids are applied as a quick dip (5 seconds) at high concentrations, or long dip (between 2 and 24 hours) at low concentrations. Water-based formulations, achieved by dissolving a hormone tablet in a specified quantity of water, are better suited to the long-dip method. The alternative alcohol-based solutions are better for the short-dip method. After dipping, allow the alcohol to evaporate by leaving the cutting to drain prior to insertion. This precludes the use of alcohol formulations on softwood cuttings which must remain turgid at all times.

Fungicides

It is worthwhile treating cuttings either with a fungicidal preparation incorporated in a hormone rooting powder or alone in solution. The latter is used as a dip for the cuttings prior to insertion, or to water them in. Fungicides are also applied to cut surfaces of root cuttings.

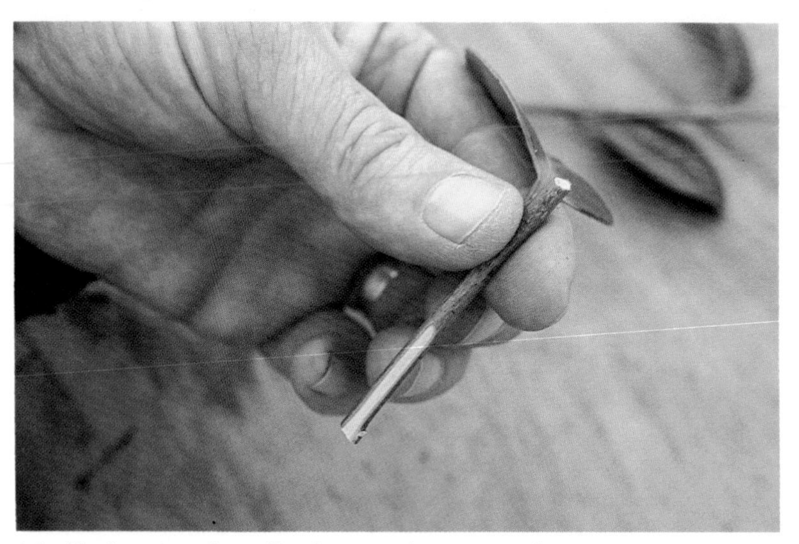
A leaf-bud cutting of camellia showing a heavy wound

Wounding

Wounding cuttings, especially those with old wood at the base, improves their ability to root, and also allows water and hormone-rooting substances to be absorbed more easily. A light wound is suitable for most cuttings; use the tip of a knife to make a 2.5 cm (1 in) slit in the bark at the base, along one or both sides. Large woody cuttings need a heavier wound. Remove a thin sliver of bark, 2.5 cm (1 in) long, along one or both sides to expose the cambium (a layer of living tissue capable of cell reproduction).

ENVIRONMENTAL FACTORS

Once the cutting has been prepared, the success of the operation is down to ensuring the environment is effectively managed and maintained. It is important to look at water, light and temperature in association with the propagation equipment you have available. Because the natural water supply has been cut off, water levels within the cutting and humidity levels in the atmosphere immediately surrounding the cutting must be carefully managed. Ideally the pH of water should be kept on the acid side of neutral, so it is advisable to use rain water if the mains water tends to be limy. Adequate light is important for leafy cuttings as it is the source of energy in photosynthesis which is necessary for root initiation and growth. Leafless, hardwood cuttings, however, depend on carbohydrate or starch levels within the cutting. Root development should keep ahead of shoot development so, ideally, artificial heat should be applied beneath the cutting so that the temperature at the base is higher than at the top.

Equipment

As a rule of thumb, the softer or less hardy the plant being propagated, the greater the level of environmental control needed for success. Site and equipment will vary considerably, from a section of the garden set aside for a nursery bed, through frames, polythene tents and tunnels, propagators and glasshouses.

Comparatively few plants will succeed from cuttings raised out of doors. Those that can include hardwood cuttings of willows (*Salix*), poplars (*Populus*) and bush fruits such as currants and gooseberries. However, certain guidelines need to be followed when preparing a site. The location should be protected and small mesh rabbit-proof fencing should be erected if necessary. In exposed places, it is worth surrounding the beds with windbreaks made out of synthetic materials or wooden laths to reduce wind speed by half.

FRAMES AND GLASSHOUSES

Frames provide a relatively inexpensive means of protection and give a surprisingly high level of success. The main disadvantage is that temperature and humidity levels fluctuate which necessitates shading, ventilation and insulation. However, if the frame is sited in a north-facing location then temperature extremes can be reduced, but rooting may take slightly longer.

There is a range of cold frames available, made from wood, aluminium alloy or brick using glass or polycarbonate sheeting. The minimum practical size is 120 × 60 cm (4 × 2 ft). The height should be 45 cm (18 in) at the back sloping to 30 cm (12 in) at the front. This will allow adequate space for containers, or rooting compost, and the height of the cuttings. The frame needs to be draughtproof so glazed panels must fit snugly. Sheets of expanded polystyrene fitted round the sides of the frame are beneficial, not only in preventing heat loss but also in building up heat within the compost. On cold nights when sharp frosts are predicted, cover the frame with hessian, old matting or similar material. If heavy frosts persist during the day, keep the covers on to maintain residual heat within the frame.

Once the rooting has taken place the frame light, or cover, should

Greenhouse bench displaying propagation aids of varying levels of sophistication – from pots clad in polythene bags to a heated propagator offering complete control over the environment

be opened increasingly wide during the day to harden off, or wean, the plants. When hardening off is complete remove the light; this will also facilitate watering and aftercare.

Frames can be equipped with soil or space heating cables. The warmth they provide will increase the range of plants which can be propagated if you are unable to run a heated propagator either in the glasshouse or your home. The cables are sold in lengths capable of heating a given area, e.g. a 75 watt cable measures 6 m (20 ft) and warms 0.75 sq m (6 sq ft). The actual area heated will depend on the temperature required. The cable is buried in moist sand – a good heat conductor – 5–8 cm (2–3 in) below the surface of the compost and arranged in loops 15 cm (6 in) apart.

Glasshouses give a greater level of protection than frames. Whatever the type of glasshouse, careful consideration needs to be given to heating, ventilation and the type of staging installed, and to what services should be supplied. As well as providing protection in their own right, they can also be used to house polythene tents, propagators and mist units.

POLYTHENE TENTS AND TUNNELS

A polythene tunnel or cloche can be used for easy-rooting deciduous shrubs, such as *Philadelphus* and *Weigela*, or for growing on young plants once rooting and hardening off are complete. Opaque polythene sheeting (150 or 300 gauge), supported by wire hoops and tensioned by wires, can be placed over prepared ground. Insert the cuttings directly into the soil, water in, then close the tunnel to maintain a humid environment.

Rigid plastic lids create a closed environment over pots or trays where small-scale propagation is being practised. Alternatively place the container in a polythene bag and support with wire hoops or canes stuck into the compost. Blow up the bag and seal with a rubber band. Polythene tents can be erected on benches within a glasshouse or on the floor, with or without bottom heat. As wide a range of cuttings can be rooted by this method as can be raised under mist (see below).

HEATED PROPAGATORS

An electrically heated propagator is suitable for using in a glasshouse or indoors, providing space and reasonable light conditions are available. Heated propagators consist of a base housing a heating cable and thermostat. A transparent cover with ventilators provides a closed environment. The amount of heat provided depends on the size of the propagator. For instance, a propagator designed to hold four seed trays will be heated with a 75 watt cable, while one taking 12 seed trays will be rated at 175 watts. Smaller domestic propagators holding two seed trays are designed for window-sill use. They are suitable for raising cuttings of easy-to-root alpines, herbaceous and shrubby plants.

Heated propagators can also be home made using soil-warming cables (see above) or heating mats which come in a range of fixed sizes.

MIST UNITS

These provide conditions conducive to rooting a wide range of plants. Their aim is to maintain a film of water droplets on the leaf surface which not only keeps humidity high but also keeps the air and leaf temperatures cool. This cooling effect means that mist units can be run in good light conditions without much need for shade. Disease does not appear to be a problem in a well-run unit as fungal spores fail to germinate. Mist units are controlled automatically by means of an electronic 'leaf' or similar device. Misting continues until enough water falls on the bare electrodes found on the electronic 'leaf' to complete an electric circuit. A solenoid valve turns off the mist until the electronic 'leaf' dries out and triggers the system again.

A mist unit enables cuttings and seedlings to root successfully and establish themselves quickly

In general, mist units are used in association with soil-warming cables (see Frames, page 10). In sunny weather the mist unit should be shaded.

If you do not possess a mist unit, place the cuttings in a closed propagator or under lightweight, opaque polythene sheeting and use a hand spray to mist them regularly.

TOOLS OF THE TRADE

The most important tools in propagation are a cuttings knife and secateurs.

Knives For most tasks a straight-bladed, carbon-steel cuttings or grafting knife is most appropriate. (Blades ground on one side only are suitable for both left and right-handed people.) Never abuse a good quality knife by cutting through string or similar – get a cheap pocket or all-purpose knife for these functions. Always clean your knife after use. The blade will need periodic sharpening on a flat carborundum stone slightly lubricated with thin oil. Never delegate this task as everyone will hold the knife at a slightly different angle. Some people prefer to use a straight-bladed scalpel for very soft material, such as dahlias, delphiniums or for small alpine plants.

The tools of the trade include cuttings and budding knives (below) kept sharp with a sharpening stone (top), and a good quality pair of secateurs

Secateurs These are employed increasingly in propagation as they are quick and easy to use. Secateurs will perform the majority of propagation work including some final cuts, especially of woody plants. Secateurs come in two basic designs, the anvil and the scissor type. The blade of anvil secateurs is ground on both sides, which cuts through the stem by crushing it against the flat anvil surface. The scissor type has one blade sharpened only on one side which slides past the unsharpened blade. The scissor type is better for propagation purposes, making a cleaner cut with less crushing of the stem than the anvil type, and is available in left and right-handed models.

Choose a pair of secateurs that feels comfortable to hold. A good quality product can be dismantled easily to allow the blade to be sharpened.

CONTAINERS

The containers used for rooting and growing on rooted cuttings depend on the number of plants being raised and personal preference. Small batches of cuttings can be rooted in 8–10 cm (3–4 in) pots. Square pots take up less space than round pots and, when used in association with soil heating cables, less heat loss occurs. Plastic pots have more or less replaced clay ones for propagation purposes as they are more hygienic, lighter, cheaper and better for using on capillary beds. Clay pots are still preferred for specific plant groups such as alpines. They are often plunged in a bed of compost or gravel as, being porous, they dry out more quickly than plastic.

Plastic seed trays 35 × 24 cm (14 × 9½ in) or half trays 24 × 17 cm (9½ × 7½ in) are more suitable for raising large numbers of cuttings. Insert the cuttings in rows and transfer them to individual containers once rooted. Polystyrene trays are also available with individual cells and are sold complete with a pusher to eject the rooted cuttings. They are useful in that the material retains heat, and the individual cells keep root disturbance to a minimum. On the down side they are bulky and difficult to clean.

Cuttings can also be rooted in soil blocks (13–50 mm/½–2 in) made with a special gadget, Jiffy 7s (expanded peat pellets held together with fine netting; 40 fill a standard seed tray), peat and paper pots and root trainers. An assortment of containers from the home can also be used so long a drainage holes are made.

LABELS AND LABELLING

Each batch of cuttings should be labelled with the name of the plant and date of insertion. It is also worth recording this information in a book, together with the donor or source, the technique used and a column for the results.

Rooting and Aftercare

COMPOSTS

The term compost is widely used in the United Kingdom to describe a growing medium for sowing seed, taking cuttings or potting up plants, as well as relating to the end product of a compost heap. In this book the term compost refers only to the growing medium. For successful propagation the compost should:

- Retain water
- Be well aerated
- Have the correct pH
- Supply both macro and micro-nutrients
- Provide anchorage

Soil-less composts

The majority of cuttings are rooted in a soil-less compost. Soil-less cutting mixtures have long been based on peat with added sharp sand, perlite or vermiculite and, to a lesser degree, rockwool, to provide a more open structure. Composted bark and coir have also been used successfully as peat alternatives. Fertilisers are added in a soluble or slow-release form, the latter often coated with resin or plastic to control their release into the compost.

The advantages of a soilless compost are that it is lightweight, clean to use, sterile and difficult to overfirm. The primary disadvantage is that once dry, it is difficult to re-wet.

Compost for propagation consists of a mixture of peat or composted bark, perlite and slow-release fertiliser

A semi-ripe cutting of *Garrya elliptica* rooted in rockwool is potted up intact

As a guide a 10-litre bag of seed and cuttings compost will fill three full-sized seed trays or 25 pots (10 cm/4 in).

Soil-based composts

Soil-based composts, of which John Innes are perhaps the best known, are suitable for potting on young plants. Alpines and root cuttings are often rooted in a soil-based compost. In the case of alpines the compost is topped off with a layer of sharp sand which provides good drainage and aeration and through which the cuttings are inserted.

John Innes (JI) composts are available under numbers denoting the level of nutrients they contain: no. 1 is suitable for potting on rooted cuttings; nos. 2 and 3 are for potting up rooted plants. These composts contain some lime, but ericaceous or lime-free versions are available for plants which do not tolerate lime, such as rhododendrons and many heathers.

John Innes composts are heavy to handle. However, it is worthwhile using it for potting on rooted cuttings, even if it is modified by adding some soil-less compost to alter the 'feel'. If soil-less composts alone are used, plants are often difficult to establish in the garden, especially on heavy soils.

In the garden

Before rooting cuttings directly in the soil, either under low polythene tunnels, cloches, or in open ground, the soil will need improving. To do this incorporate organic matter in the autumn. If the soil is heavy clay or not in a good, friable condition grit will also need to be added.

FEEDING AND WATERING

For healthy growth plants require about 16 different elements which are described as either macro-nutrients (nitrogen, phosphate and potash) or micro-nutrients, often referred to as trace elements. If there is either an imbalance or shortage of nutrients, plants will display symptoms of this. A complete fertiliser will always state the proportions of nitrogen, phosphate and potash (N, P, K). Proprietary fertilisers should give application rates as these will vary according to the formulation. Supplementary feeding of rooted cuttings will start in early spring, as soon as growth commences, and should become a weekly routine. If the compost is dry, moisten it thoroughly before feeding.

Incorrect watering is by far the most common cause of failure when attempting to raise young plants. There is no real substitute for experience in being able to 'read' your plants' requirements and to assess whether they are too wet, too dry or just right. Water with care, especially during late autumn and winter when overwatering is most likely to occur.

Whether using a watering can or hand lance attached to a hose, always attach a rose. This will ensure water is applied directly but gently to the compost surface, so the structure of the compost is maintained and compaction is minimised.

There is a range of watering systems available for plants that have been potted on: capillary matting, trickle irrigation, seep hoses, overhead spray lines. Choose a system that is appropriate for the plants you are propagating and the numbers involved.

HARDENING OFF

Plants rooted under mist, in closed cases or in a cold frame will need to be hardened off prior to planting out. If they have been growing under mist they should be kept in the mist unit until after they have been potted up. Once established, young plants can be gradually hardened off, or weaned, by removing to an open bench shaded from direct sun and with, at first, an air temperature similar to that in the mist unit. Spray over plants manually at intervals decreasing in frequency as each day passes. Cuttings raised in a closed case or cold frame, will need gradually increased levels of ventilation. For frames, start by giving daytime ventilation and then wedging the frame open overnight providing there is no risk of frost.

Taking Cuttings

There are many techniques available for taking cuttings from various parts of a plant in order to create an identical plant. However, variations of technique and timing specific to particular plants are listed in the charts on pages 52–64. The following parts of a plant can be used for cuttings:

- Root
- Stem: softwood, greenwood, semi-ripe, hardwood
- Stems: layering
- Leaf

Many plants can be propagated in more than one way. You should adopt the method you are most comfortable with, given the material available.

Romneya coulteri is reliably raised from root cuttings

Clump of Oriental poppies (*Papaver orientale*) lifted and showing roots suitable for cuttings

ROOT CUTTINGS

Apart from plants that sucker naturally it is not often clear which plants will regenerate from root cuttings. The presence of a thick, fleshy root, one containing adequate carbohydrate levels, is a good indication. Grafted plants should be avoided as root cuttings will regenerate the rootstock and not the desirable grafted cultivar.

Timing The most appropriate time to take root cuttings is during the dormant period prior to commencement of growth. In the majority of instances this will be mid to late winter. However, with early-flowering herbaceous subjects such as Oriental poppies (*Papaver orientale*), root cuttings can be taken in late summer.

Method

Select young, healthy, vigorous plants up to three years old. Lift the plant, wash off surplus soil to make root selection easier and also to reduce the risk of pest and disease. Once suitable roots have been severed, trim back the remaining roots and top growth of the parent to stimulate fresh growth and replant.

Select roots of pencil thickness growing fairly close to the surface; in the case of some herbaceous plants (*Phlox*, *Primula denticulata* and *Romneya*) they will be thinner. Length of the cuttings can vary from 2.5 cm (1 in) to 10–15 cm (4–6 in). In the majority of instances root cuttings are inserted vertically so the polarity of the cutting must be noted, particularly if more than one cutting is taken from a single root.

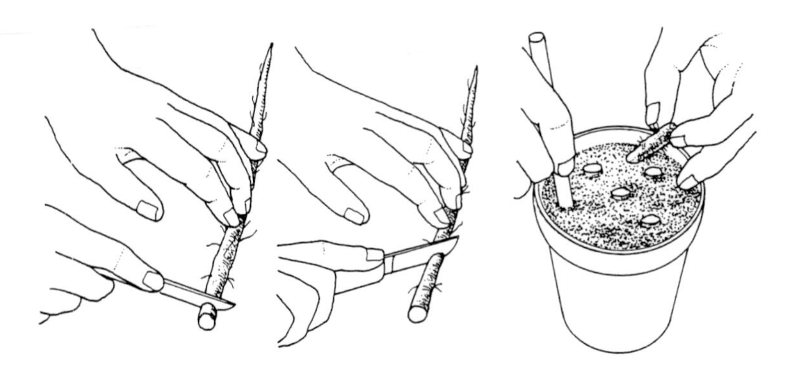

1 The top of the cutting, the end nearest the crown where the growing shoots will appear, is cut straight across
2 A slanting cut is made at the end furthest from the crown. Treat the cut surfaces with a proprietary fungicide to prevent disease
3 Insert cuttings so the top is just below the compost surface

Root cuttings can be raised in open ground, a cold frame or glasshouse. Always label with the name of the plant and the date of insertion.

Open ground is only suitable for easy-rooting trees and shrubs, including *Ailanthus*, *Rhus typhina* and *Chaenomeles*. Choose a protected site where the soil has been regularly cultivated. A V-shaped trench about 5 cm (2 in) deeper than the length of the cutting should be dug with one side vertical. If the soil is on the heavy side then cover the base of the trench with sharp sand. Position cuttings vertically, 10 cm (4 in) apart, so the top is just below the soil surface. Backfill with friable soil. Shoots may not appear for four months or more.

A frame or low-level polythene tent or tunnel can be used to raise a wider range of plants including: alpines (*Erodium*, *Geranium*), herbaceous plants (*Acanthus*, *Anchusa*, *Papaver*), shrubs (*Clerodendrum trichotomum*) and trees (*Catalpa*, *Paulownia*). The soil should be well cultivated in the autumn and covered with plastic sheeting so that soil temperatures and moisture levels will be conducive to rooting. Insert cuttings vertically in rows using a dibber. The top of the cutting should be just beneath the surface. Cover the surface with 6 mm ($\frac{1}{4}$ in) washed lime-free grit to reduce capping. (This is not essential in a cold frame.) Distances between cuttings and rows will vary according to the plant. Cuttings should shoot within two or three months although some subjects may take up to a year.

Small numbers of root cuttings, and those of certain alpines (*Anemone*, *Pulsatilla*) and herbaceous plants (*Limonium*, *Primula*) are better planted in pots placed in a cold frame. Fill the container with JI potting compost no. 1 or soilless equivalent. After inserting the cuttings vertically apply a top dressing of 6 mm (¼ in) lime-free grit.

A glasshouse will further widen the range of plants which can be successfully raised from root cuttings, including the difficult *Romneya coulteri*. Take cuttings 2.5–5 cm (1–2 in) long and plant in the same way as described for the cold frame (see above). If basal heat of 18 °C (65 °F) is available they will shoot in four to six weeks. Keep air temperature as low as possible but above freezing. Pot up plants during the early summer, as soon as they have been hardened off.

STEM CUTTINGS

These are the most common types of cuttings and can be divided into four groups:

- Softwood
- Greenwood
- Semi-ripe
- Hardwood

SOFTWOOD CUTTINGS

An extremely broad range of primarily deciduous plants can be propagated from softwood cuttings, and the younger the wood the greater the capacity to develop roots. However, the survival rate of the cutting declines rapidly once it has been detached from the parent plant. The young leaves are still expanding and lose water very quickly, so steps to avoid wilting need to be taken.

Timing Softwood cuttings are generally taken during early spring as soon as the first flushes of new growth appear. However, glasshouse subjects (including evergreens) can be taken at most times of the year. It is essential to take and insert cuttings while they are still turgid, preferably early in the morning. Have ready all your equipment – boxes, pots, compost – so any delay between collection and insertion is minimised. If the parent plant is any distance from the cuttings bench, collect the cuttings into a polythene bag that is moist inside. Partially inflate the bag to avoid

Taking softwood cuttings

1 Collect cuttings (here *Penstemon*) into a polythene bag

2 Take cuttings below a node and remove the lower leaves

3 Dip base of cuttings into hormone rooting powder containing fungicide

4 Insert cuttings into compost with the aid of a dibber

5 A propagating case will provide a closed environment so the cuttings will not dry out

6 As soon as the cuttings have formed a root system, pot them up individually

bruising the cuttings and seal. Place it in a large, white bag further to reduce moisture loss caused by the sun's heat. If you are unable to deal with the cuttings immediately, place them in the fridge until you are ready.

Method

Most stem cuttings are taken below a node or leaf joint. Cuttings of shrubby plants have a better chance of success if they are taken with a heel (a sliver of the previous year's wood). Internodal cuttings of easy-to-root subjects such as fuchsias and forsythias can also be taken. The size of the cutting will vary but is normally between 5–10 cm (2–4 in). If the cutting is any longer, remove the growing tip. This redistributes the plant hormones throughout the cutting and often improves its rooting capability.

Prepare cuttings out of direct sunlight. Remove the lower leaves with a sharp knife or scalpel. In general make a straight cut across the stem, however, a slanting cut to trim the heel is also acceptable.

Dip the base of the cutting in a fungicidal solution or powder, or hormone rooting powder no. 1 which contains a fungicide. The compost used can vary from a proprietary cuttings compost to 100 per cent sharp sand or perlite (a volcanic rock which has been expanded by heating). Fill the pot or tray with compost, firming down at the corners, and level with a scraper board. Insert and lightly firm in the cuttings with a dibber. Label, and water – this will also help firm in the cuttings. Place the cuttings as quickly as possible under mist, a closed case or a low-level polythene tent.

It is important to remove any dying or diseased leaves on a regular basis and to spray cuttings with a fungicide once a week. Rooting can take place in as little as one week and such quick-rooting material can then be hardened off and potted up. However, softwood cuttings of woody plants will take longer (azaleas taking six weeks or more). In these instances cuttings should be retained in their containers and given a liquid feed at regular intervals, potting on the following spring.

GREENWOOD CUTTINGS

These cuttings are taken between the softwood and semi-ripe stages. The stem is harder than in a softwood cutting and the stems and leaves are darker with age. The majority of deciduous shrubs are propagated by greenwood cuttings, as well as pelargoniums (old-fashioned bedding geraniums).

Timing By implication greenwood cuttings are taken during late spring and mid-summer, although glasshouse subjects can be rooted at other times of the year from appropriate material. Cuttings taken from current season's wood are collected in the same way as softwood cuttings.

Method

Cuttings of shrubs are taken with a heel of older wood. This will reduce the incidence of stem rot which may develop on nodal stem cuttings. Greenwood cuttings are generally longer than softwood cuttings, from 8–13 cm (3–5 in). Large leaves can be reduced in size so they take up less space and also lose less moisture through the leaf surface. A light cut or wound (see page 9) is beneficial, followed by a fungicidal spray or drench.

A number of plants, especially climbers, grow particularly vigorously. In such cases nodal cuttings may prove difficult given the length of stem between the leaves, and internodal cuttings are more suitable. Internodal cuttings are taken approximately 5 cm (2 in) below a node. The top of the cutting is trimmed directly above a node as excess stem may rot and die back to this point.

1

2

Internodal cuttings
1 Cuttings of *Clematis armandii* prepared by trimming just above a node and about 3–4 cm ($1\frac{1}{2}$ in) below a node, and reducing the leaf area significantly
2 Wound the stem and dip in hormone rooting powder
3 Insert cuttings in cuttings compost covered with a layer of grit

3

When the cuttings have been prepared, dip the base of each in hormone rooting powder no. 2 and treat as for softwood cuttings (see page 22). Once rooted, keep cuttings in their containers and give a liquid feed at regular intervals during the growing period. Pot them on during the following spring. Some greenwood cuttings root readily (such as fuchsias, forsythias and *Phygelius*) and these can be potted up during the late summer.

SEMI-RIPE CUTTINGS

During the late summer cuttings from evergreen plants, including conifers, can start to be taken. Stems are more woody than greenwood cuttings, and winter buds will probably have developed.

Timing This will vary from late summer through to late winter. Select material from the current season's growth. The chart on pages 52–64 gives guidance on the best times to take semi-ripe cuttings for a selection of plants. Some evergreens will root better before frosts e.g. *Ilex*, whereas others, such as *Osmanthus*, root more readily taken after a cold period. Semi-ripe cuttings of glasshouse plants can be taken whenever appropriate material is available. Semi-ripe cuttings, as they are more woody than greenwood cuttings, are better adapted to withstanding moisture loss. Collect material in the same way as for softwood cuttings (see page 22).

Method

Take nodal cuttings, preferably with a heel to reduce the incidence of stem rot, 8–13 cm (3–5 in) long. The length will vary depending on the plant. Heathers, for example, will be no longer than 5 cm (2 in), while cuttings of some rhododendrons and other large

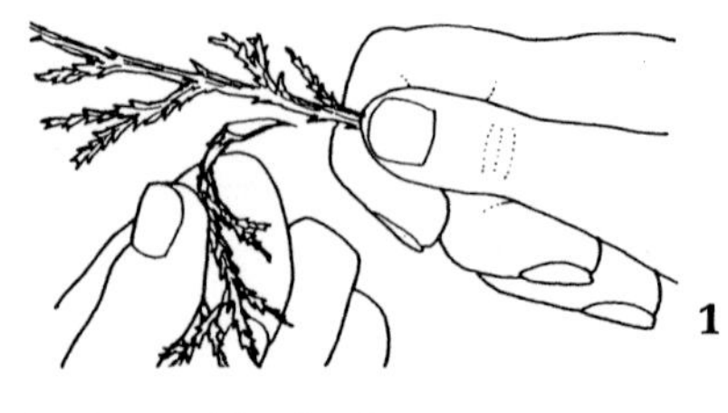

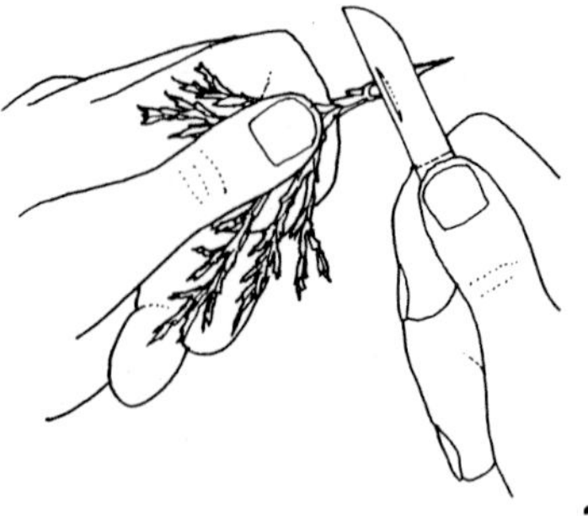

Cuttings with a heel
1 Pulling off a vigorous young shoot with a 'heel' of older wood
2 Trim the heel before dipping in a hormone rooting compound

shrubs can be up to 15 cm (6 in). Cuttings with large leaves can be cut in half so they take up less space and moisture loss is reduced. Treat cuttings with a fungicide, either by dipping or spraying. Wounding is beneficial (see page 9).

Select material with care, especially when propagating conifers. For instance, both adult and juvenile foliage can be found on many *Chamaecyparis* cultivars. The juvenile growth may also grow horizontally rather than vertically. Upright growth showing a regular pattern of apical (tip) buds must be selected if vertical growth is to be achieved. This also applies when propagating pines (*Pinus*), spruces (*Picea*) and firs (*Abies*) by grafting.

Once the cuttings have been prepared, dip them in hormone rooting powder nos. 2 or 3 depending on the size of the cutting: for heathers (*Erica*, *Calluna* and *Daboecia*), *Hebe* and *Cistus* use no. 2; for rhododendrons, hollies (*Ilex*) and *Elaeagnus* use no. 3.

A variety of methods can be used to root semi-ripe cuttings depending on the facilities available and the plant's ability to root. A cold frame allows you to raise a wide range of evergreens providing you are not expecting fast results. Cuttings are normally inserted up until the end of September when outside temperatures are still buoyant. The frame can be lined with polystyrene to raise temperatures at the point of rooting.

With heat A heated frame, or a propagator or mist unit within the glasshouse provides controlled conditions for cuttings. They are either inserted into an open bed of compost (satisfactory if large quantities of similar plants are being rooted) or into seed trays or other modules. When using a heated frame cover the cuttings with a white lightweight polythene sheet to maintain a humid atmosphere around the cutting. Remove once or twice a day in order to ventilate the frame, remove dropped leaves (a source of fungal infection) and get rid of condensation which would otherwise drip on the compost. (Over-wet compost will adversely affect the cuttings' ability to root.) Cuttings rooted in a mist unit, propagator or heated frame should not be potted up until the following spring, after hardening off.

Without heat Cuttings will take longest to root in a cold frame. Keep the frame closed and covered with hessian or matting during frosty periods. Once the cuttings have rooted they can be hardened off, prior to potting up. Any cuttings not rooted but callused over should be replaced after paring away the base of the callus, feed them during summer and, providing they have rooted, pot up the following spring.

LEAF-BUD CUTTINGS

Evergreen shrubs can be raised from leaf-bud cuttings if cutting material is in short supply or a plant needs bulking up quickly. The technique is particularly appropriate for propagating camellias, mahonias and *Ficus*.

1

Leaf-bud cuttings
1 Select the current season's growth (here mahonia). Cut the stem about 5 cm (2 in) below a bud and again directly above it. If the buds are alternate make the cut slant away from the bud
2 Where the leaf is large, as with mahonias, reduce it by half. Insert a split cane to support the cutting if required
3 Mahonia cutting inserted into rooting medium

2

3

HARDWOOD CUTTINGS

This is a very simple technique of plant raising requiring only the most basic facilities for success. A limited range of trees including willows (*Salix*) and poplars (*Populus*), and shrubs (*Cornus alba*, *Philadelphus* and various bush fruits) can be propagated by this method using vigorous one-year-old wood that is fully ripened.

Timing Hardwood cuttings can be taken throughout the dormant period. Those taken immediately after leaf-fall or at the end of the dormant period, just before the leaves open, are most likely to succeed.

Method

The production of vigorous, one-year-old shoots can be encouraged by stooling or hard pruning plants the previous year. The length of the cutting varies considerably: from 2.5 cm (1 in) in the case of vine eyes to 1.8 m (6 ft) for willow sets. The average cutting is of pencil thickness, with about six buds and measures approximately 20 cm (8 in) (the length of the average pair of secateurs).

Usually all buds are retained on the cutting. However, with gooseberries and red currants which are grown on a 'leg', or where suckering needs to be discouraged, as in the case of rose rootstocks, the lower buds are rubbed out retaining the top three buds only.

Hardwood cuttings
1 Trim straight across the base of the cutting, the end nearer the roots (here *Cornus stolonifera* 'Flaviramea')
2 Make a slanting cut at the top end, just above a node
3 A selection of hardwood cuttings have been bundled, dipped in hormone root powder and plunged in sharp sand prior to lining out in spring
4 Hardwood cuttings inserted in a trench with vertical sides. Sharp sand in the bottom ensures good drainage and encourages rooting down

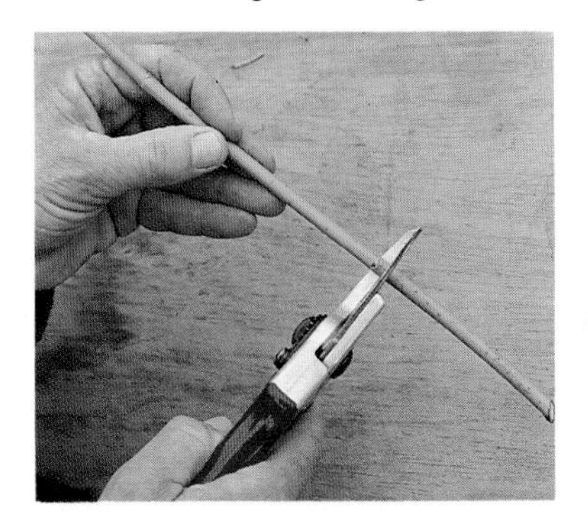

1

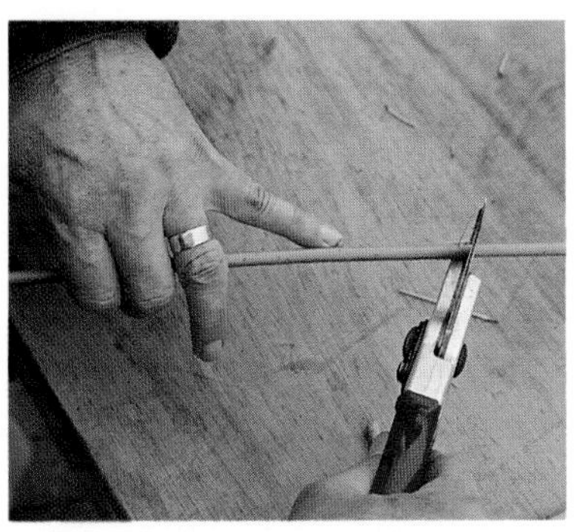

2

3

4

5 Alternatively, hardwood cuttings (here *Ficus*) can be raised in pots. Illustrated are the mature stem, prepared cuttings, equipment for the job including hormone rooting powder, and pot of cuttings ready for the cold frame

---

To ensure cuttings are inserted the correct way up, the top is trimmed at an angle immediately above the bud, while the lower end is cut straight across. It is not necessary to make the base cut immediately below a node unless the plant has a hollow or pithy stem like *Sambucus*.

Once the cuttings have been prepared, dip them in no. 3 hormone rooting powder. If cuttings were taken at leaf-fall, bundle up to ten of them together – any more and the central cuttings will desiccate. Insert the bundles in a large pot of sharp sand placed adjacent to a cold frame or in a sheltered, shady part of the garden, so that the tops are only just visible. Water in. The advantage of bundling over direct planting is that conditions will not fluctuate as wildly as they can do in the garden. Just before growth commences in the spring, the cuttings can be untied and treated as follows.

Prepare the ground, if necessary, the previous autumn so the soil is light and friable. Take out a vertical trench sufficiently deep so only the top 2.5–5 cm (1–2 in) of the cutting is exposed above the surface. If the ground is heavy incorporate some sharp sand into the base of the trench. Space the cuttings 10–15 cm (4–6 in) apart with 30–38 cm (12–15 in) between the rows. Backfill and firm the soil around the cuttings. After frost the soil may need re-firming.

If single-stemmed trees are being raised, then the entire cutting should be buried so the top bud is level with the soil surface.

Follow this method for raising cuttings in a cold frame leaving 10 cm (4 in) between the rows or inserting them in pots.

Rooted cuttings can be lifted and planted out or potted on by the following autumn.

Vine eyes

These are taken from dormant one-year-old shoots that have been cut into 2.5–4 cm (1–1½ in) lengths each containing one bud. Although they are much smaller they are treated as hardwood cuttings. Cuttings can be made in one of two ways:

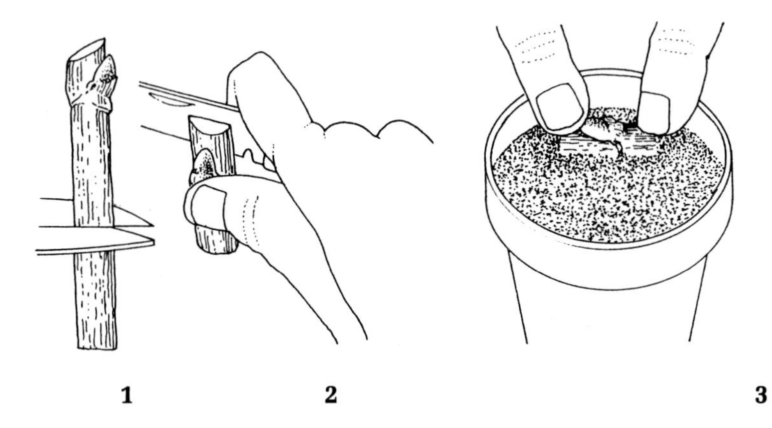

1 If the cuttings are inserted vertically, trim them to leave a bud at the top end, just proud of the surface of the compost
2 If the cuttings are laid horizontally then cut the stem in half longitudinally and trim so the bud is central
3 The finished cuttings are laid on top of the compost, bud uppermost and just above the surface of the compost

Stand the pots or trays of cuttings on the glasshouse bench or in a propagator. Water in the cuttings but do not overwater subsequently. Cuttings raised in seed trays can be potted individually later on in the season.

An alternative method is to insert a single stem of the previous year's growth, still attached to the parent, through the base of an empty pot. Fill the pot with compost so a clear section of stem, up to 30 cm (12 in), is visible above the surface. The rod will produce roots and grow, eventually producing bunches of grapes. At this time the rod can be severed from the parent plant and taken to the table complete as an edible decoration.

Willow sets

Cut 1.8 m (6 ft) lengths from the current season's wood, just before growth starts in spring. Insert these sets directly where they are to grow for an 'instant' shelter belt or arbour. Rooting is improved by making a longer wound than normal (see page 9).

LEAF CUTTINGS

A few plants can be raised by leaf cuttings. These are mainly house plants belonging to the begonia and gesneriad families (the latter includes African violets) and peperomias.

Timing The cuttings can be taken throughout the year, although rooting is slower during the winter months. Leaf cuttings will start losing moisture as soon as they are severed from the parent plant, so have ready a pot or seed tray filled with cuttings compost and lightly firmed. Insert cuttings into the compost using a dibber. Cuttings are inserted at a shallow angle so the leaf is virtually lying on top of the compost with the cut stem just under the surface. The most suitable environment for rooting leaf cuttings is a heated closed case maintaining a temperature of between 18–21°C (65–70°F). High levels of humidity are required and the cuttings should be shaded from direct sunlight. It is important to maintain good hygiene by spraying with a fungicide and ensuring clean cuts are made.

Leaf-stem (petiole) cuttings

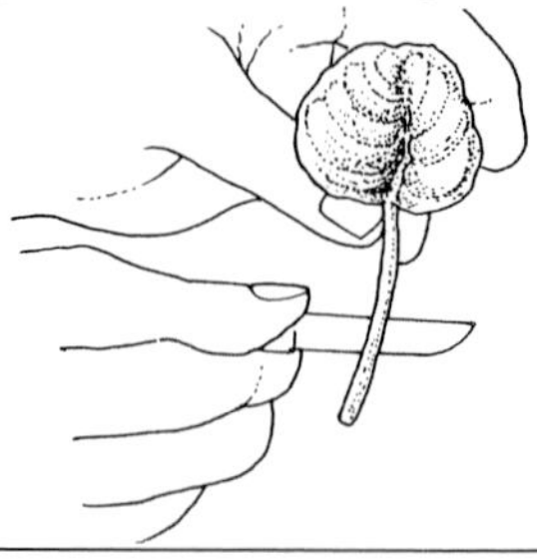

Select a fully-expanded leaf and cut it from the parent plant complete with leaf stem. Reduce the stalk to 5 cm (2 in), making a clean cut with sharp knife or scalpel

Label and water in with a fungicidal solution. Maintain good hygiene by removing any leaves showing signs of disease. The cuttings will produce new plants in six to eight weeks, and these can be potted up soon after. Cut off the parent leaf once the young plant is established.

Midrib cuttings

This method is used for leaves which possess a thick central midrib, in particular *Streptocarpus*. A leaf is cut into sections which are inserted into a tray of cuttings compost. Make sure the polarity of the leaf is correct, with the cut closest to the leaf base in the compost. If you are in doubt about the polarity of the sections, make them V-shaped following the line of the lateral veins which join the midrib roughly at an angle of 45 degrees. Like lead petiole cuttings, the sections should be inserted shallowly. Rooting takes place in six to eight weeks with the new plant forming on the cut midrib.

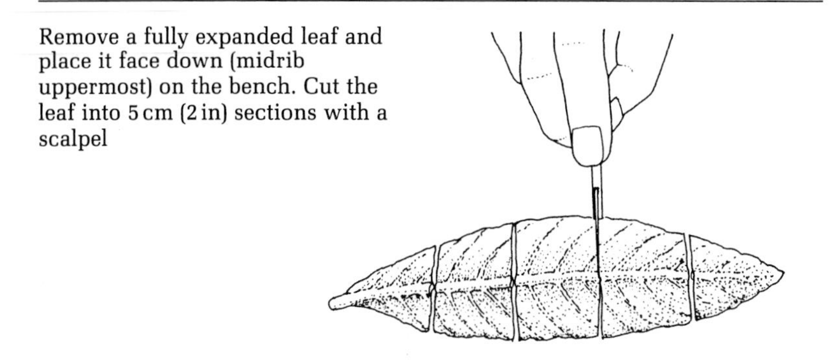

Remove a fully expanded leaf and place it face down (midrib uppermost) on the bench. Cut the leaf into 5 cm (2 in) sections with a scalpel

One or two monocotyledonous plants such as *Sansevieria* and *Lachenalia* can be rooted by this method. However, the variegated *Sansevieria trifasciata* 'Laurentii' should be increased by division. Plants raised from leaf cuttings will have all-green foliage and lack the attractive variegation.

Leaf squares

Begonia rex can be propagated by either leaf squares or leaf slashing. Remove an undamaged, mature leaf and turn it upside down, exposing the veins. Cut the leaf into 2.5 cm (1 in) squares each containing a major vein. Either lay the squares on the surface of the compost or insert them in the compost as described for midrib cuttings ensuring a major vein is inserted into the compost. Rooting takes place in about six weeks.

Alternative method using a whole leaf

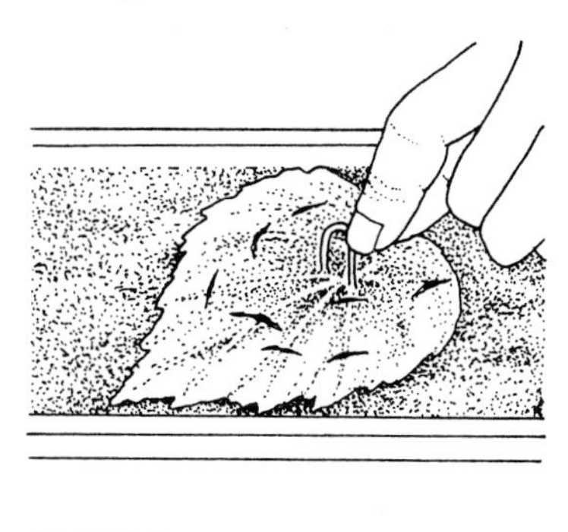

The leaf is laid face down and the main veins are slashed with a series of scalpel cuts. Lay the leaf on the surface of the compost – veins down – and anchor with staples

Left: *Tolmiea menziesii*, the pick-a-back plant, develops foliar embryos
Right: *Asplenium bulbiferum* also produces tiny plantlets on its foliage. These can be detached and grown on

Foliar embryos

A few plants have the ability to produce plantlets spontaneously from groups of cells or foliar embryos found at specific locations on the leaves. *Tolmiea menziesii* develops plantlets at the base of its leaf surface where it joins the petiole. *Kalanchoe diagremontiana* produces plantlets around the margins of the leaf. In both cases plantlets can be separated from the parent and potted up.

In the case of ferns such as *Asplenium bulbiferum*, the fronds bearing foliar embryos are cut from the parent plant, pinned flat on a seed tray of cuttings compost and watered in.

Other plants need stimulation to produce foliar embryos. *Kalanchoe blossfeldiana* and *Echeveria* will develop foliar embryos on leaves once they are detached from the parent plant. Place the leaves on sand or cuttings compost in a seed tray and keep fairly dry.

Layering

Layering encourages the development of roots on a stem while it is still attached to the parent plant. Rooting can be achieved without the array of propagation aids often needed for cuttings. In many instances those plants that are difficult to root can be propagated by layering and success can be achieved by most people regardless of experience.

Tips for success

It is important to constrict the stem for rooting to take place. There are several ways this can be done. The shoot can be bent or twisted, alternatively it can be partially cut through. With water and nutrients still being supplied to the layered shoot, the preparation time is not critical. Covering the developing shoot so it is in the dark stimulates adventitious roots to develop where the stem is lengthening.

Ideally the site should be flat and sheltered. The quality of the soil is important: those that are heavy, hot and dry soils in summer or compacted all inhibit rooting. Therefore, before layering, fork over the soil, removing any weeds and incorporating cuttings compost to improve its moisture-holding capacity and aeration. The addition of a slow-release fertiliser is also beneficial.

Loganberries, like many other *Rubus* species and hybrids, can be tip layered

TIP LAYERING

This technique is used to raise plants of the genus *Rubus*, particularly the fruiting varieties such as blackberry, loganberry and tayberry, which must be grown from virus-free stock. Rooting occurs when the tip of the current season's growth is held in contact with moist soil.

During the early part of the summer when growth has reached 60 cm (2 ft), the selected shoot tip is pinched out to promote the development of side shoots. When these have reached 90–150 cm (3–5 ft), arch them over and mark the ground where the tips touch. Incorporate some cuttings compost into the soil and prepare an L-shaped trench, 10 cm (4 in) deep. The length of the trench will depend on the layers. Place the growing tips at the base of the vertical side so the tips are pointing upwards and pin them into position with wire staples. Cover the growing tips with soil, gently firm and water in.

During the winter rooted layers can be carefully lifted, avoiding too much damage to fibrous roots. Cut them close to the point of rooting and pot up, or plant out in a nursery bed, for a season. The length of shoot between the parent plant and the layer is shortened back to 10–15 cm (4–6 in) from the base of the parent plant.

SIMPLE LAYERING

Simple layering occurs when shoots, ideally one-year-old shoots, growing close to the ground are bent vertical with soil or compost covering the point when the stem is bent. Many woody plants, both deciduous and evergreen, respond to this treatment and it is commonly practised on plants that are difficult to root from cuttings or where one or two plants only are needed.

Shoots growing close to the ground may need encouraging, so cut back the plant the previous year to produce suitable vigorous growth. Prepare the ground as described on page 35 and take out an L-shaped trench, 10–15 cm (4–6 in) deep, positioned so that 15–30 cm (6–12 in) of the layered stem will be visible above the ground. Start layering in early spring just before growth commences, or autumn.

In general, rooting will have taken place by autumn when deciduous plants can be lifted and planted out or potted on. Evergreens are better lifted the following spring. Some difficult-to-root subjects may need to remain attached to the parent plant for a further 12 months. In these cases, apply a slow-release fertiliser and top dressing of cuttings compost in spring.

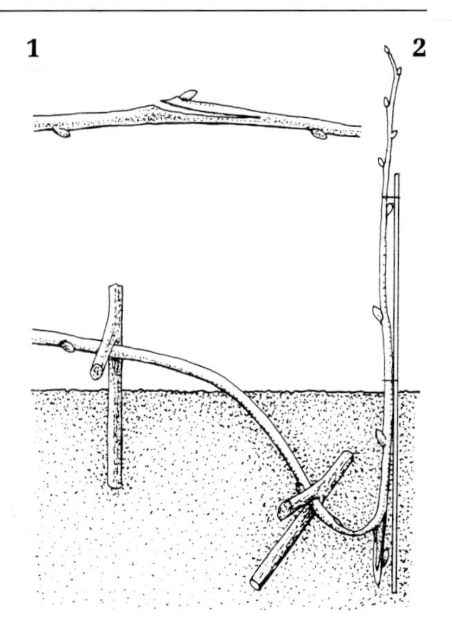

1 Constrict the stem either by twisting, making a cut above about 2.5 cm (1 in) long along the stem or girdling, by removing completely a section of bark about 5 mm ($\frac{1}{4}$ in) wide from around the stem. Dust hormone rooting powder no. 3 over the cut surface
2 Peg down the layer in the bottom of the trench. Push in a split cane adjacent to the vertical side of the trench. Tie shoot to cane. For evergreens, cut off leaves below ground level. Fill in the trench with soil or compost and carefully firm around the layer. Water in, and continue to water as necessary throughout the summer. Sever the rooted layer from the parent plant. Trim back the severed stem to within 10–15 cm (4–6 in) of the main stem

Compound or serpentine layering

This modification of simple layering is suitable for climbing plants that are difficult to root, such as *Lapageria* and *Wisteria*.

Select vigorous shoots in the spring. Starting fairly close to the parent plant, dig a hole up to 10 cm (4 in) deep with a trowel or hand fork. Wound stem and insert a section of stem, including wound and a node, in the hole and pin down with a staple. Backfill with a mixture of soil and cuttings compost. Loop the next section of stem overground before digging another hole and pinning down the stem at a node. Repeat these steps so alternate nodes are buried along the length of the stem. Rooted layers are lifted the following winter or late spring and potted on. Stem sections can be anchored into pots sunk in the soil rather than into the ground and severed when rooted.

AIR LAYERING

This technique, also known as Chinese layerage, encourages aerial parts of the plants to root. In most instances it is used to raise glasshouse plants that require high levels of humidity for successful rooting, such as *Ficus* and *Citrus*. It is carried out in spring on the previous season's growth or in autumn on wood ripened during the same season.

Left: With its long flexible stems *Lapageria rosea* is a suitable candidate for serpentine layering
Right: Heathers that have grown straggly can be propagated by dropping, a form of layering

The stem is girdled by removing a section of bark, up to 12 mm (½ in) wide, from around the stem at a point between 15–30 cm (6–12 in) behind the growing point. Alternatively a longitudinal cut, about 2.5 cm (1 in) long, can be made towards the growing point. Dust the cut surface with hormone rooting powder no. 2 or 3. Where a longitudinal cut has been made, a matchstick is inserted to keep the cut surfaces apart, and sphagnum moss packed in between to ensure rooting takes place along the length of the cut. A white polythene bag (measuring 15–20 cm/6–8 in) is slit at the base and placed over the stem. Seal it below the girdled or cut surface with waterproof tape. Pack moist, but not wet, sphagnum moss in the polythene bag so that it surrounds the cut surface and stem. Seal the bag at the top.

High temperatures inside the bag can be a problem, 21 °C (70 °F) is the optimum and temperatures much in excess of this will inhibit rooting. Rooting can occur in three months but more usually a complete growing season is needed for sufficient roots to have formed to sustain the plant. The layer is then cut below the roots. Remove the polythene bag and tease out the sphagnum moss prior to potting up. Cut back any snags on the parent plant.

DROPPING

This is a method of layering low-growing evergreen shrubs, such as heathers and dwarf rhododendrons, that have become straggly. It

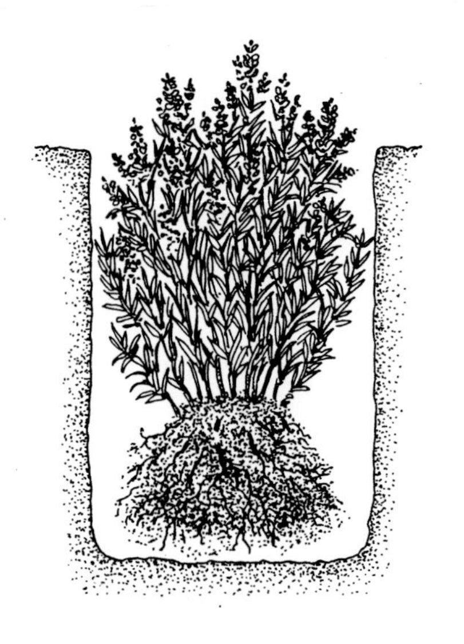

1

2

Dropping
1 Put the lifted plant into the planting pit
2 Backfill by hand so that only the growing tips remain exposed

is generally carried out in spring though on heavy clay soils autumn is a more appropriate time. The selected plant will need to be transplanted. Prepare the site so the soil is friable and of high fertility and, ideally, incorporate some cuttings compost. Dig out a planting pit large enough to take the plant and deep enough so that only the growing tips will be visible above the ground. Carefully lift the plant, prune out any dead wood or damaged roots and place it in the waiting pit. (With heathers it may also be advisable to remove some of the growth so there are reasonable gaps between the exposed tips.) Backfill by hand, working the soil/cuttings compost mixture round the stems. Firm with your fingers as you backfill. Once planting is completed, label and water in.

Rooting usually takes place during the first or second season. To check, carefully draw away the compost during the following autumn or spring and, if rooted, sever the layers, and discard the parent plant.

Division

The majority of herbaceous plants, including some alpines, and a number of woody plants are increased by division.

HERBACEOUS PLANTS

In propagation terms, herbaceous plants are divided into three groups:

- Those with fibrous crowns
- Those with compacted crowns
- Those of a semi-woody nature

Division is the simplest way of making more of your herbaceous perennials, including hostas, tiarella and alchemilla

Timing

This depends on when a plant flowers. A general rule of thumb is the later a plant flowers, the later it is divided. Thus those plants that flower early in the year are divided in late summer; mid-season flowerers are divided between autumn and spring; and autumn-flowering plants in spring. However, if plants are demanding or difficult to grow, it is advisable to divide them immediately before growth commences in the spring to give them every chance of establishing themselves.

Plants with fibrous crowns

These include *Aster*, *Doronicum*, hardy geraniums, *Hemerocallis*, *Phlox*, *Rudbeckia*, *Veronica*. These plants are easily multiplied by splitting up the crowns in early winter or spring. Having cut down old flowering stems and any dead and dying growth, the plant is lifted and excess soil removed. If the clump is to be divided into small pieces, wash the roots in a bucket of water. Then break or cut off sections from the edges of the crown, as this is where the more

1

2

3

Plants with fibrous crowns
1 Lift the crown (here *Hemerocallis*, day lily)
2 If a limited number of larger portions are needed, divide the original clump by inserting two garden forks or border forks back to back and drawing the handles together
3 Before replanting clean up the divisions by cutting away with secateurs any damaged roots and shoots. Discard the woody central portion

vigorous shoots will be found. The older woody centre can be discarded.

Replant the divided sections in ground prepared in advance and plant the divisions at the same depth as before, carefully working in soil among the fibrous roots. Firm and water in. Alternatively, pot up the divisions into a proprietary potting compost, water in and place in a cold frame. This will enable young plants to establish without bringing them on too quickly.

Plants with tough compacted crowns

These include *Helleborus*, *Hosta*, *Sedum spectabile*. Crowns that are too dense to pull apart are lifted in spring just before growth begins. Cut the crown into sections, each piece containing at least one shoot and adequate roots. These sections are either potted up or planted. Small portions benefit from a dusting with fungicide before planting.

Plants that flower very early in the spring such as *Helleborus orientalis*, as best divided during autumn when root growth begins.

Bergenia, lifted to show offsets which can be detached to form new plants

Semi-woody herbaceous plants

Phormium and *Cortaderia* are two examples of this type of plant. They are lifted in the spring. Because the crowns are so tough, spades or even old axes are called into play. Select young material from the margins and drastically cut back the top growth, especially in the case of *Phormium*, to minimise moisture loss through the leaves.

ALPINES

A number of alpines can be multiplied by division. Those that flower in spring, such as *Arabis* and saxifrages, are divided

Left: Strawberry plants showing runners, long stems that develop plantlets at intervals
Right: *Aesculus parviflora* is an example of a suckering shrub which can be divided

immediately after flowering. Autumn-flowering alpines such as *Gentiana sino-ornata* and *Saxifraga fortunei* are also divided in the spring but before growth commences.

OFFSETS AND RUNNERS

An offset is a plantlet that arises close to the crown, either above or below ground. Sempervivums produce offsets in the spring that do not root until later in the year.

A runner is a stem that creeps overground producing plantlets along its length. Strawberry and *Ajuga reptans* produce runners that root at every node. The resulting plantlets are lifted and potted up or planted out in the same season they arise, otherwise a mat of growth quicky develops.

SUCKERS

Woody plants that spread by suckers can also be divided. *Aronia* and *Aesculus parviflora* are two examples, but avoid any plants that have been grafted as only the rootstock will be perpetuated. Lift the sucker in the spring, cutting it cleanly from the parent plant. The top growth is also reduced by about a third (more if a large number of suckers are lifted) to cut down moisture loss and to encourage new growth.

Tiny bulbils growing in the leaf axils of *Lilium bulbiferum*

Bulbs and other Storage Organs

Bulbs, corms, tubers, tuberous roots, rhizomes and pseudobulbs are all storage organs. They can be increased by division in various ways according to type.

BULBS

Bulbs consist of fleshy scale leaves each with an auxillary bud which are arranged concentrically around the growing point. They are divided into two groups: tunicate and non-tunicate. Tunicate bulbs, which include narcissus and tulips, have a dry membraneous outer covering. Non-tunicate or scaly bulbs, such as lilies and fritillarias, are fleshy and lack this protective covering so they are more easily damaged and susceptible to drying out.

Another difference is that the roots of tunicate bulbs start to form at the beginning of the growing season, whereas those of non-tunicate bulbs develop during late spring and summer and persist through to the following year.

All bulbs increase naturally by division, either by producing bulblets, bulbils or simply by producing a new bulb during the growing season. In certain instances, particularly bulbous irises, the original plant produces a large number of small bulbs which take several years to build up to flowering size. There are also various techniques which can be employed to induce bulbs to increase.

Bulbils

These are produced in the axils of the leaves of some lily species such as *Lilium bulbiferum* and *L. lancifolium*. Others, including *L. candidum* and *L. × testaceum* can be induced to form bulbils by removing the flower buds.

Collect bulbils in late summer and plant them directly into pots for growing on. Push the bulbils into firmed soil-less potting compost so the top half of the bulbil is exposed. Cover them with grit, label, water and place in a cold frame. During the following season water and feed with a proprietary liquid feed before planting them out during autumn. They should flower two or three years after collection.

Bulblets

Unlike bulbils, bulblets are produced underground. They form naturally on some lilies such as *L. longiflorum*, on other lilies they can be induced. This is an efficient way of multiplying disease-free or virus-free stocks.

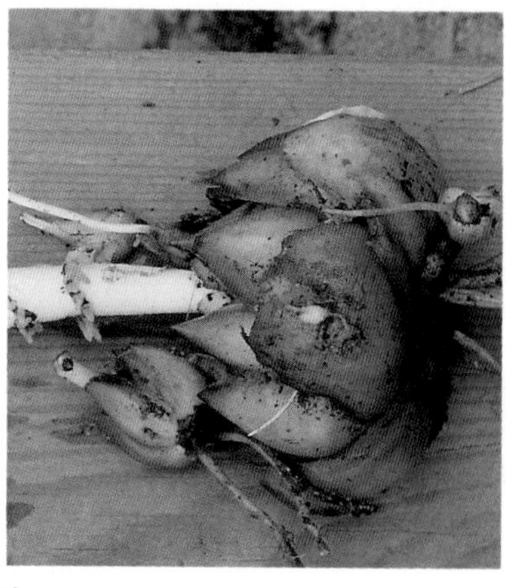

1

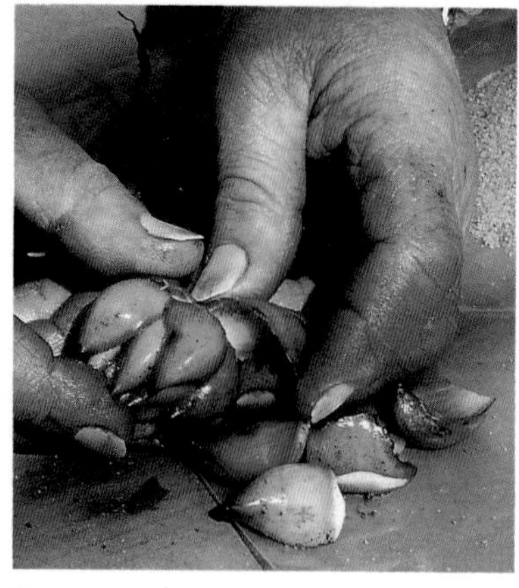

2

Scaling lilies
1 After flowering, during late summer, a bulb (here *Lilium speciosum*) is carefully lifted, cleaned and the outer layer of scales removed.
2 The scales can be simply snapped close to the basal plate or cut off with a sharp knife. Insert scales vertically into a pot or seed tray filled with cuttings compost and water in using a fungicidal solution. Place the pot or tray in a propagator maintained at 18–21 °C (65–70 °F). Replant the mother bulb. Pot them up and transfer to a cold frame after hardening off. The bulblets are grown on for a further year before planting out.

An alternative method is to mix the scales with moist vermiculite and place in a polythene bag. Inflate the bag, seal and place in a warm, dark place such as an airing cupboard, at 21 °C (70 °F) for three to five weeks. After this time bulblets should have formed at the base of the scales. Carefully remove the scales from the vermiculite and pot up in cuttings compost, covering the bulblets. Gradually harden off and keep in a frost-free environment for the first winter.

Chipping

This technique is used to multiply tunicate bulbs such as narcissus, snowdrops (*Galanthus*) and nerines and others that produce tightly packed scale leaves. Select bulbs in midsummer and clean thoroughly.

Using a scalpel or thin-bladed budding knife, cut each bulb in half from top to bottom. Each half is then subdivided further making sure each section has a piece of the basal plate. (Commercially each bulb is cut into 16 chips.) Dip the chips in a fungicidal solution, drain and mix with moist vermiculite in the ratio of two to three parts vermiculite to one of chipped bulbs. Put the mixture into a polythene bag, inflate, seal and place in an airing cupboard for about 12 weeks.

Check at regular intervals and discard any that show signs of rot. Once the bulblets have formed plant them in pots or trays, harden off and place in a frost-free environment for a year. They will flower in three or four years.

Scooping and scoring

These are techniques used to increase bulbs that are slow to form bulblets, such as hyacinths.

Scooping is performed on mature bulbs, 17 cm (7 in) or more in circumference which have previously been lifted, dried and cleaned. Scoop out the entire basal plate with a curved scalpel, short-bladed knife or sharp teaspoon, to make a concave depression (about a quarter of the way into the bulb). Dust the cut with fungicide and place, cut side up, on a wire mesh tray or dry sand in an airing cupboard at 21–25°C (70–77°F). Callus will develop over the cut. However to prevent desiccation sand will need to be moistened periodically.

After about a week remove the bulbs from the airing cupboard. Bulblets will form at the base of the scales in about three months. As soon as they are visible pot up the mother bulb, still inverted, in soilless compost so that the bulblets are just below the surface. Top dress with grit. Gradually harden off, keeping the bulbs in a frost-free environment over the first winter. During the following summer the bulblets will grow while the mother bulb disintegrates. Once the foliage dies down, pot up the bulblets individually. They will take a further three years to reach flowering size.

Scoring A technique very similar to scooping except it differs in the preparation of the mother bulb. Make two or three incisions across the basal plate and treat as described for scooping, above. Fewer but larger bulblets will result from this treatment as fewer scale leaves have been cut.

CORMS

Although a corm, crocus for example, appears similar to a bulb, it is, in fact, a swollen stem. The dry, thin membranes enclosing the corm protect it against injury and desiccation. Each year a new corm develops around the base of the old corm. However, if greater numbers are required a large healthy corm can be cut vertically into three to six sections at planting time. Dust the cut surfaces with a proprietary fungicide and leave in a warm, well-ventilated space at 21–30°C (70–85°F) for about 48 hours to allow cut surfaces to callus over. Pot up in a well-drained potting compost by pushing the corm pieces into the surface and covering with grit.

Cormels

These are miniature corms which are produced between the old and new corms. Numbers of cormels vary, with gladiolus producing up to 50 per corm. Lift the corms in the autumn and split off the cormels. If this proves difficult, place the mother corm in a warm room for about 12 hours and after that time they should split off easily.

Mix the cormels with slightly moistened vermiculite and place in a polythene bag. Inflate the bag, seal and store over winter in a frost-free environment at 5°C (40°F). Plant out the cormels in the spring.

Alternatively space the cormels in a pot filled with compost to which extra grit has been added, to ensure it is free draining, and cover them with grit. Keep the compost dry over winter and start watering in spring. Cormels usually take two years to flower.

TUBERS

A tuber is an underground swollen stem that develops during the growing season, remains dormant over the winter and produces new shoots in the spring. The 'eyes' on a tuber's surface are in fact nodes, each consisting of small buds. The potato is the best-known example of a tuber, but caladiums, cyclamen and water-lilies (*Nymphaea*) are also tuberous.

Tubers are increased by cutting them into sections with a sharp knife, just before growth starts in spring. Each section must have at least one bud or 'eye'. Dust the cut surfaces with fungicide and place the sections in a warm, dry place (21°C/70°F) for two days, before planting them out at twice their depth.

Tubercles

Miniature tubers are found in the auxillary buds of a few plants such as *Achimenes* and *Dioscorea batatas*. These are collected in the autumn, stored dry over winter and planted in the spring.

Left: Gladiolus corm showing cormels developing between the new and old corm
Right: *Gladiolus* 'The Bride' is propagated from young cormels

Tuberous roots

To the casual observer tuberous roots look like tubers. However they lack nodes and internodes. The buds are produced only at the crown or stem end. Tuberous roots like those of dahlia can be increased by division, making sure each piece has at least one bud. Other techniques produce more plants, such as stem and leaf cuttings.

Lift crowns towards the end of October, clean and dry (by arranging them upside down on a rack). After dusting with fungicide, store tubers in dry compost or wrap in newspaper and place in a frost-free environment (4–10°C/39–50°F). During early spring crown buds will develop showing where it is most appropriate to cut. Dust cuts with fungicide and leave to dry for 24 hours prior to potting in an open compost. Keep on the dry side until shoots begin to grow.

PSEUDOBULBS

These swollen fleshy stems are made up of one or a number of nodes and occur widely among orchids. Pseudobulbs can be increased by offsets as in *Dendrobium* or by division (*Miltonia* and *Odontoglossum*). Divide immediately before growth begins by

Left: Dahlia tubers lifted in autumn and stored over winter will sprout in spring to produce good cutting material. Alternatively the tuberous roots can be cut into pieces, each bearing a bud
Right: Cymbidium orchids can be divided into sections of three or more pseudobulbs and potted up to make more plants

removing sections containing three or more pseudobulbs. In the case of *Cymbidium*, pull or cut off back bulbs (older swollen stems surplus to the plant's requirements) so each group retains some root system. Pot up the divisions in a proprietary orchid compost and keep on the dry side. A new shoot will appear within a month or two of potting.

RHIZOMES

These specialised perennial stems, which grow horizontally just below the soil surface, are divided either at the beginning of the growing season or immediately after flowering. This is done according to whether their growth habit is:

- Thick and fleshy, as in rhizomatous irises;
- Crown rhizome, a much-branched clump as developed in peony and rhubarb;
- Long and slender, as displayed by mint (*Mentha*) and lily-of-the-valley (*Convallaria majalis*).

Thick and fleshy rhizomatous irises

These are divided immediately after flowering in early to mid-summer. Cut out the old part of the rhizome leaving the current year's growths which will be approximately 10 cm (4 in) in length. The root system is cut back to 8–10 cm (3–4 in) and the foliage to 15 cm (6 in). This reduces moisture loss while the root system becomes established.

1

2

1 Lift clumps of rhizomatous irises after flowering
2 Make a clean vertical cut with a sharp knife immediately behind the leaf scar
3 Plant the rhizome in a shallow trench so the top is visible above the soil, and water in

3
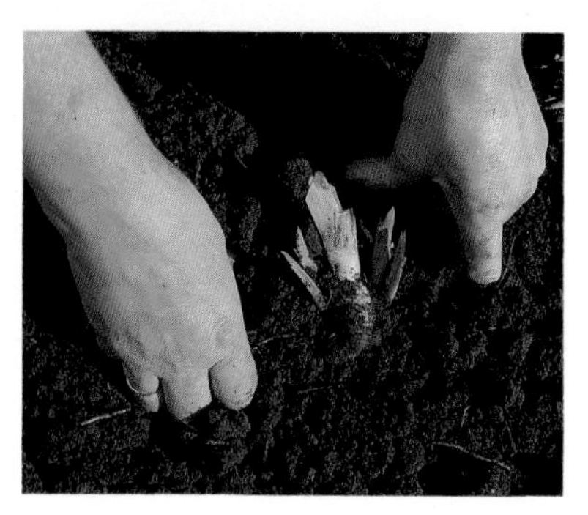

Crown rhizomes

These are lifted in spring, just before growth starts. After teasing or washing off the soil, chop the crown into a few large pieces with a spade. Trim the cut surface smooth with a garden knife, before replanting the pieces into prepared ground at the same level they were before lifting.

However, if more plants are required, place the cleaned clump into a box containing moist peat, coir or vermiculite and put in a frost-free shed. As the shoots start to grow, they can be removed with a piece of the parent crown. Each section must have at least one bud. Dust the cut surfaces with fungicide. Lay the divisions on a tray and keep in a warm place (21°C/70°F) for 24 hours to callus over before potting them up into cuttings compost to which extra sand, grit or perlite has been added.

Long slender rhizomes

Divide these just before growth begins in spring. Because of their rapid growth rate even small divisions can be planted out directly.

Propagation Chart

AN INDEX TO TECHNIQUES

Although not all the plants listed below have a specific text reference, the appropriate technique is explained on the given page(s).

AYR = all year round

Plant	Timing (months)	Technique	Page
Abelia	6–7	greenwood cuttings	24
Abutilon	6–7	greenwood cuttings	24
Acanthus	6–8	leaf cuttings (frame, propagator)	32
	12–2	root cuttings	20
Achillea	2–4	division	41
Achimenes	12–3	division, rhizomes	48
Aesculus parviflora	11–2	division, suckers	43
		root cuttings (open ground)	20
Ailanthus	12–2	root cuttings (open ground)	20
Ajuga reptans	2–4	division	41
	6–8	runners	43
Alchemilla (lady's mantle)	2–4, 10–12	division	41
Allium (ornamental onions)	6–8	bulbils	45
Amelanchier	10–11, 2–4	layering	35
	6–8	greenwood cuttings	24
Anchusa: herbaceous, alpines	7–9, 2–4	root cuttings (frame, propagator)	20
herbaceous only	2–4	division	41
Anemone: herbaceous	2–4	division	41
rhizomatous	7–9	division (pots frame)	51
Arabis	8–10	division	42
	6–7	softwood cuttings	22
Argyranthemum	7–9	greenwood cuttings	24
Armeria (thrift)	2–4	division	41
Aronia	11–2	division, suckers	43
	2–3	root cuttings (frame, open ground)	20
Artemisia: perennials	2–4	division	41
sub-shrubs	3–4, 8–9	semi-ripe cuttings	26

Begonia rex is propagated from whole leaf cuttings or by leaf squares

Plant	Timing (months)	Technique	Page
Asplenium bulbiferum	AYR	foliar embryo from mature fronds	34
Aster (Michaelmas daisies)	2–4	division	41
Aubretia	6–8	softwood cuttings	22
Bamboos	2–4	division	42
Begonia: cane-type	AYR	softwood cuttings	22
	2–4	division of tuberous roots	49
B. rex	AYR	leaf squares, whole leaf	33
Berberis: deciduous	6–8	greenwood cuttings	24
	10–11	hardwood cuttings	28
evergreen	8–10	semi-ripe cuttings	26
Bergenia	8–9	division	42
Black currants	10–11	hardwood cuttings	28
Blackberry and hybrids	6–8	leaf-bud cuttings	28
	6–8	tip layer	36
Brachyglottis	8–10	semi-ripe cuttings	26
Buddleja	10–11	hardwood cuttings	28
	6–8	greenwood cuttings	24
Buxus (box)	8–10	semi-ripe cuttings	26
Caladium	2–4	division, tubers (propagator)	48
Calluna	7–10	semi-ripe cuttings with heel	26
	2–4	dropping	38

Plant	Timing (months)	Technique	Page
Camellia	7–10	leaf-bud cuttings	28
	7–10	semi-ripe cuttings	26
Campanula: alpines	4–6	softwood cuttings	22
herbaceous	2–4	division	41
Campsis	2–3	root cuttings (frame, propagator)	20
Canna	2–4	division, rhizomes	51
Caryopteris	6–7	greenwood cuttings	24
Catalpa	10–2	root cuttings (frame, propagator)	20
Ceanothus: deciduous	6–8	greenwood cuttings	24
evergreen	8–10	semi-ripe cuttings	26
Ceratostigma	6–7	greenwood cuttings	24
Chaenomeles	10–2	root cuttings (open ground, frame, poly tunnel)	20
	6–8	greenwood cuttings	24
Chamaecyparis	8–10	semi-ripe cuttings	26
Choisya (Mexican orange blossom)	8–10	semi-ripe cuttings	26
Chrysanthemums see *Dendranthema*			
Cistus	7–9	semi-ripe cuttings	26
Citrus	2–4	air layering	37
	7–8	greenwood cuttings	24
Clematis	7–9	compound layering	37
	6–8	greenwood internodal cuttings	25
Clerodendrum trichotomum	10–2	root cuttings (frame, poly tunnel)	20
Coleus see *Solenostemon*			
Convallaria majalis (lily of the valley)	10–4	division, rhizomes (open ground)	51
Cornus alba	10–11, 2–4	hardwood (open ground, frame)	28
Cornus florida, C. kousa	2–4	simple layering	36
Cortaderia	2–4	division	42
Corylus (hazel)	10–2	simple layering	36
Cotinus coggygria	6–7	greenwood cuttings	24
Cotoneaster: deciduous	6–7	greenwood cuttings	24
evergreen	8–10	semi-ripe cuttings	26
Crocosmia (montbretia)	12–4	division, corms	47

Clerodendrum trichotomum var. *fargesii* is raised from root cuttings

Plant	Timing (months)	Technique	Page
Crocus	6–8	division, corms	47
Cyclamen: autumn-flowering	2–3	division, tubers	48
spring-flowering	7–9	division, tubers	48
Cytisus	6–7	greenwood cuttings	24
Daboecia	7–10	semi-ripe heel	26
Dahlia	2–3	division, tuberous roots	49
	2–3	softwood cuttings	22
Daphne: deciduous	6–7	greenwood cuttings	24
evergreen	8–10	semi-ripe cuttings	26
Delphinium	2–3	softwood cuttings	22
	2–4	division	41
Dendranthema (chrysanthemums)			
perennials	2–4	division	41
florists cvs.	2–3	softwood cuttings	22
Dendrobium	5–6	division, pseudobulbs	49
Deutzia	6–7	greenwood cuttings	24
Dianthus: pinks, alpines	7–9	greenwood cuttings (pipings)	24, 57
florist carnations	AYR	greenwood cuttings (pipings)	24, 57

Plant	Timing (months)	Technique	Page
Diascia	7–9	softwood cuttings	22
	2–4	division	41
Dicentra	8–10	division	41
Dioscorea batatas	2–4	division, tubercles	48
Doronicum	4–5, 8–9	division	41
Echeveria	8–9	foliar embryo	34
Echinops	2–4	division	41
Elaeagnus	8–10	semi-ripe cuttings	26
Enkianthus	6–8	greenwood cuttings	24
Epimedium	8–9	division	41
Erica: excluding tree heathers	2–4	dropping	38
Erica: spp. and cvs.	7–10	semi-ripe cuttings with a heel	26
Erodium	9–11	root cuttings (frame)	20
	2–3	stem cuttings	22
Eryngium	10–2	root cuttings	20
	2–4	division	41
Escallonia	8–10	semi-ripe cuttings	26
Euphorbia: herbaceous	2–4	division	41
shrubs	8–10	semi-ripe cuttings	26
Ferns: hardy	2–4	division	41
Ficus benjamina	AYR	leaf-bud cuttings	28
		semi-ripe cuttings	26
Ficus elastica	AYR	air layering	37
		leaf-bud cuttings	28
Forsythia	6–7	greenwood cuttings	24
Fragaria (strawberry)	7–9	division, runners	43
Fuchsia	3–9	softwood cuttings	22
Galanthus (snowdrop)	6–7	division, chipping	42
Garrya	8–10	semi-ripe cuttings	26
Gaultheria	8–10	semi-ripe cuttings	26
Gentiana:			
autumn-flowering	3–4	division	43
		softwood cuttings	22
spring-flowering	3–4	division	43
	6–7	stem cuttings	20
Geranium: alpines	1–3	stem cuttings	20
herbaceous	2–4	division	20
Grasses	2–4	division	41, 42

Cuttings known as 'pipings' of alpine dianthus ready for insertion

Plant	Timing (months)	Technique	Page
Gladiolus	1–4	division, cormels	48
Gooseberries	10–11	hardwood cuttings	28
Hebe	8–10	semi-ripe cuttings	24
Hedera (ivy)	8–10	leaf-bud cuttings	26
	8–10	semi-ripe cuttings	24
Helenium	2–4	division	41
Helianthemum	3–4, 8–10	semi-ripe cuttings	24
Helianthus	2–4	division	41
Helichrysum: alpines, herbaceous	2–4	division	41
shrubs	8–10	semi-ripe cuttings	26
Helleborus orientalis	8–9	division	42
Hemerocallis (day lily)	2–4	division	41
Hosta	2–4	division	42
Humulus (hop)	5–7	simple layering	36
	5–7	compound layering	37
	5	softwood cuttings	22
Hyacinthus (hyacinth)	6–7	scooping, scoring	47

Plant	Timing (months)	Technique	Page
Hydrangea	6–8	greenwood cuttings	24
Hypericum	6–8	greenwood cuttings	24
Ilex (holly): deciduous	6–7	greenwood cuttings	24
evergreen	8–10	semi-ripe cuttings	26
Iris: bulbous	6–8	bulblets	46
rhizomatous	5–7	divide rhizomes after flowering	50
Jasminum (jasmine)	6–7	greenwood cuttings	22
Juniperus (juniper)	8–10	semi-ripe cuttings	24
Kalanchoe blossfeldiana	7–9	foliar embryo (propagator)	34
K. diagremontiana	7–9	foliar embryo (propagator)	34
Kniphofia (red hot poker)	2–4	division	41
Kolkwitzia	6–8	greenwood cuttings	24
Lachenalia	2–3	leaf midrib	32
	8–9	bulblets	46
Lapageria	2–4	compound layering	37
Lavandula (lavender)	3–4, 7–9	semi-ripe cuttings	26
Laurus (bay)	8–10	semi-ripe cuttings	26
Lavatera (mallow): herbaceous	7–9	greenwood cuttings	24
shrubs	9–11	hardwood cuttings	28
Leucojum (snowflakes)			
spring flowering	6–8	scooping and scoring	47
summer flowering	2–4	scooping and scoring	47
Ligustrum (privet)	7–9	semi-ripe cuttings	26
Lilium (lily)	8–10	bulbils	44
	8–10	bulblets scaling	46
Limonium (sea lavender)	10–11, 2–3	root cuttings (frame, open ground)	20
Lobelia: perennials	8–9	leaf-bud cuttings of flowering stem	28
	2–4	division	41
Lonicera (honeysuckle):			
deciduous	6–8	greenwood cuttings	24
evergreen	8–10	semi-ripe cuttings	26
Magnolia: deciduous	6–7	greenwood cuttings	24
evergreen	2–3	simple layering	36
	8–10	semi-ripe cuttings	26
Mahonia	8–10	leaf-bud cuttings	28
Mentha (mint)	2–4	division, rhizomes	51

Evergreen hollies, such as *Ilex* × *altaclerensis* 'Lawsoniana', are propagated from semi-ripe cuttings

Plant	**Timing (months)**	**Technique**	**Page**
Metasequoia (dawn redwood)	6–8	greenwood cuttings	24
	10–11	hardwood cuttings	28
Miltonia	5–6	division, pseudobulbs	49
Miscanthus	2–4	division	41, 42
Monarda	2–4	division	41
Myrtus (myrtle)	8–10	semi-ripe cuttings	26
Narcissus (daffodil)	8–9	division, bulbs	44
	6–7	chipping	46
Nerine	8	division, bulbs	44
	5–6	chipping	46
Nymphea (water lily)	2–4	division, tubers	48
Odontoglossum	5–6	division, pseudobulbs	49
Origanum	5–6	softwood cuttings	22
	2–4	division	41, 42
Osmanthus	10–12	semi-ripe cuttings	26
Osteospermum	7–9	greenwood cuttings	24

Plant	Timing (months)	Technique	Page
Paeonia (peony): herbaceous	2–4	division, crown rhizome	51
Papaver (poppy)	8–10	root cuttings (frame, open ground)	20
Parahebe	8–10	semi-ripe cuttings	26
Parthenocissus	5–6	softwood cuttings	22
	7–8	greenwood cuttings	24
Passiflora (passion flower)	7–8	greenwood cuttings	24
Paulownia	10–2	root cuttings (frame, propagator)	20
Pelargonium	8–9	greenwood cuttings	24
Penstemon: alpines, half-hardy	7–9	greenwood cuttings	24
Peperomia	AYR	leaf-petiole cuttings	32
Perovskia	4–5	softwood cuttings with a heel	22
	7–9	greenwood cuttings	24
Philadelphus (mock orange)	7–8	greenwood cuttings	24
	10–12	hardwood cuttings	28
Photinia: deciduous	7–8	greenwood cuttings	24
evergreen	8–10	semi-ripe cuttings	26
Phlox: herbaceous	2–4	division	41
	10–11, 2–3	root cuttings (frame, propagator)	20
alpines	6–8	greenwood cuttings	24
Phormium	2–4	division	42
Phygelius	6–8	greenwood cuttings	24
Pittosporum	8–10	semi-ripe cuttings	26
Polemonium	2–4	division	41
Polygonatum (Solomon's seal)	2–4	division, rhizomes	50
Populus (poplar)	10–2	hardwood cuttings	28
Potato	1–3	division, tubers	48
Potentilla: herbaceous	2–4	division	41
shrubs	6–8	greenwood cuttings	24
Primula (primroses and polyanthus)	5–8	division, timing depends on species	41
Primula denticulata	8–10, 2–3	root cuttings (frame, propagator)	20
Prunus Laurocerasus	8–10	semi-ripe cuttings	26
Pulmonaria (lungwort)	4–5, 8–9	division	41
Pulsatilla (Pasque flower)	9–11	root cuttings (pots frame)	20
Pyracantha (firethorn)	7–10	semi-ripe cuttings	26

Raise red currants from hardwood cuttings, rubbing out all but the top three buds so plants will grow on a leg

Plant		Timing (months)	Technique	Page
Red currants		9–11	hardwood cuttings	28
Rhododendron:	dwarf	2–4	dropping	38
	deciduous	5–6	softwood cuttings	22
	evergreen	8–10	semi-ripe cuttings with a heel	26
Rhubarb		2–3	division of crown rhizome	51
Rhus typhina		11–12, 2–3	root cuttings (open ground)	20
		11–2	division, suckers	43
Ribes (flowering currants)		6–8	greenwood cuttings	24
		10–11	hardwood cuttings	28
Romneya coulteri		3–4	root cuttings (greenhouse, propagator)	20
Rosa		10–11, 2–3	hardwood cuttings	28
Rosmarinus (rosemary)		7–9	semi-ripe cuttings	26

Divide *Sedum spectabile* in spring

Plant	**Timing (months)**	**Technique**	**Page**
Rubus (bramble)	6–8	greenwood cuttings	24
	6–8	leaf-bud cuttings	28
	6–8	tip layer	36
Rudbeckia	2–4	division	41
Saintpaulia (African violet)	AYR	leaf-petiole cuttings	32
Salix (willow)	10–4	hardwood sets	31
Salvia (sage): sub-shrubs	7–9	greenwood cuttings	24
perennials	2–4	division	41
Santolina	3–4	softwood cuttings with a heel	22
	8–9	semi-ripe cuttings	26
Sambucus (elder)	10–4	hardwood cuttings	28

Saintpaulia (African violet) can be increased from leaf petiole cuttings

Plant	Timing (months)	Technique	Page
Sansevieria	AYR	midrib cuttings	32
	AYR	leaf section	33
Sansevieria trifasciata 'Laurentii'	2–4	division, offsets	43
Sarcococca	8–10	semi-ripe cuttings	26
	3–4	division, suckers	43
Satureja (winter savory)	7–9	greenwood cuttings	24
	2–4	division	41
Saxifraga: alpines	3–5	division	42
S. stolonifera	8–9	runner	43
Schizostylis	2–4	division, bulbs	44
Sedum spectabile	2–4	division	42
Sempervivum	2–4	division, offsets	43
Senecio	2–4	division	41
Sinningia (gloxinia)	2–4	division, tubers	48

Plant	**Timing (months)**	**Technique**	**Page**
Solenostemon (coleus)	2–4	softwood cuttings	22
	8–9	greenwood cuttings	24
Spiraea	6–8	greenwood cuttings	24
	10–11, 2–3	hardwood cuttings	28
Stachys	2–4	division	41
Strawberry see *Fragaria*			
Streptocarpus	AYR	midrib cuttings	32
	AYR	leaf sections	33
Taxus (yew)	8–10	semi-ripe cuttings	26
Teucrium	8–10	semi-ripe cuttings	26
Thymus (thyme)	6–10	greenwood cuttings	24
	2–3	dropping	38
Thuja	8–10	semi-ripe cuttings	26
Tiarella	2–4	division	41
Tolmiea menziesii (pick-a-back plant)	7–9	foliar embryo	34
Tradescantia	2–4	division	41
tender species	AYR	greenwood cuttings	24
Tropaeolum tuberosum	2–4	division, tubers	48
Verbascum: herbaceous	2–4	division	41
shrubs	10–11, 2–3	root cuttings	20
	3–4	semi-ripe cuttings	26
Verbena	8–9	greenwood cuttings	24
Veronica	2–4	division	41
Viburnum: deciduous	6–8	greenwood cuttings	24
evergreen	8–10	semi-ripe cuttings	26
Vinca (periwinkle)	7–10	semi-ripe cuttings	26
	2–4	division	43
Vitis (vines, grapes)	12–1	vine eyes	31
Weigela	7–8	semi-ripe cuttings	26
Wisteria	6–7	compound layering	37
Zantedeschia	2–4, 8–9	division, rhizomes	51